The Penal Cluster

by

Randall Garrett

Double 9 BOOKS

The Penal Cluster
by Randall Garrett

ISBN: 978-93-59326-48-1

Published by

DOUBLE 9 BOOKS
2/13-B, Ansari Road
Daryaganj, New Delhi – 110002
info@double9books.com
www.double9books.com
Tel. 011-40042856

ABOUT THE AUTHOR

Phillip, Gordon Randall David Garrett (December 16, 1927 – December 31, 1987) was a science fiction and fantasy author from the United States. In the 1950s and 1960s, he contributed to Astounding and other science fiction periodicals. He taught Robert Silverberg how to market enormous amounts of action-adventure science fiction and worked with him on two novels about Earthmen upsetting a peaceful agrarian civilisation on an extraterrestrial planet. Garrett is best known for the Lord Darcy books, which include the novel Too Many Magicians and two short story collections set in an alternate world where a joint Anglo-French empire led by a Plantagenet dynasty has survived into the twentieth century and magic works and has been scientifically codified. The Darcy books are full of jokes, puns, and references (specially to works of detective and spy fiction: Lord Darcy is fashioned after Sherlock Holmes), with elements reappear frequently in the detective's lesser works. Michael Kurland went on to write two more Lord Darcy novels. Garrett used several pen names, including David Gordon, John Gordon, Darrel T. Langart (an anagram of his name), Alexander Blade, Richard Greer, Ivar Jorgensen, Clyde Mitchell, Leonard G. Spencer, S. M. Tenneshaw, and Gerald Vance. As "Randall of Hightower" (a pun on "garret"), he was also a founding member of the Society for Creative Anachronism.

Tomorrow's technocracy will produce more and more things for better living. It will produce other things, also; among them, criminals too despicable to live on this earth. Too abominable to breathe our free air.

THE clipped British voice said, in David Houston's ear, *I'm quite sure he's one. He's cashing a check for a thousand pounds. Keep him under surveillance.*

Houston didn't look up immediately. He simply stood there in the lobby of the big London bank, filling out a deposit slip at one of the long, high desks. When he had finished, he picked up the slip and headed towards the teller's cage.

Ahead of him, standing at the window, was a tall, impeccably dressed, aristocratic-looking man with graying hair.

"The man in the tweeds?" Houston whispered. His voice was so low that it was inaudible a foot away, and his lips scarcely moved. But the sensitive microphone in his collar picked up the voice and relayed it to the man behind the teller's wicket.

That's him, said the tiny speaker hidden in Houston's ear. *The fine-looking chap in the tweeds and bowler.*

"Got him," whispered Houston.

He didn't go anywhere near the man in the bowler and tweeds; instead, he went to a window several feet away.

"Deposit," he said, handing the slip to the man on the other side of the partition. While the teller went through the motions of putting the deposit through the robot accounting machine, David Houston kept his ears open.

"How did you want the thousand, sir?" asked the teller in the next wicket.

"Ten pound notes, if you please," said the graying man. "I think a hundred notes will go into my brief case easily enough." He chuckled, as though he'd made a clever witticism.

"Yes, sir," said the clerk, smiling.

Houston whispered into his microphone again. "Who is the guy?"

On the other side of the partition, George Meredith, a small, unimposing-looking man, sat at a desk marked: MR. MEREDITH—ACCOUNTING DEPT. He looked as though he were paying no attention whatever to anything going on at the various windows, but he, too, had a microphone at his throat and a hidden pickup in his ear.

At Houston's question, he whispered: "That's Sir Lewis Huntley. The check's good, of course. Poor fellow."

"Yeah," whispered Houston, "if he is what we think he is."

"I'm fairly certain," Meredith replied. "Sir Lewis isn't the type of fellow to draw that much in cash. At the present rate of exchange, that's worth three thousand seven hundred and fifty dollars American. Sir Lewis might carry a hundred pounds as pocket-money, but never a thousand."

Houston and Meredith were a good thirty feet from each other, and neither looked at the other. Unless a bystander had equipment to tune in on the special scrambled wavelength they were using, that bystander would never know they were holding a conversation.

"... nine-fifty, nine-sixty, nine-seventy, nine-eighty, nine-ninety, a thousand pounds," said the clerk who was taking care of Sir Lewis's check. "Would you count that to make sure, sir?"

"Certainly. Ten, twenty, thirty, ..."

While the baronet was double-checking the amount, David Houston glanced at him. Sir Lewis looked perfectly calm and unhurried, as though he were doing something perfectly legal—which, in a way, he was. And, in another way, he most definitely was not, if George Meredith's suspicions were correct.

"Your receipt, sir." It was the teller at Houston's own window.

Houston took the receipt, thanked the teller, and walked toward the broad front doors of the bank.

"George," he whispered into the throat mike, "has Sir Lewis noticed me?"

"Hasn't so much as looked at you," Meredith answered. "Good hunting."

"Thanks."

As Houston stepped outside the bank, he casually dropped one hand into a coat pocket and turned a small knob on his radio control box. "Houston to HQ," he whispered.

"London HQ; what is it, Houston?" asked the earpiece.

"Leadenhall Street Post. Meredith thinks he's spotted one. Sir Lewis Huntley."

"Righto. We've got men in that part of the city now. We'll have a network posted within five minutes. Can you hold onto him that long?"

Houston looked around. Leadenhall Street was full of people, and the visibility was low. "I'll have to tail him pretty closely," Houston said. "Your damned English fogs don't give a man much chance to see anything."

There was a chuckle from the earphone. "Cheer up, Yank; you should have seen it back before 1968. When atomic power replaced coal and oil, our fogs became a devil of a lot cleaner."

The voice was quite clear; at the London headquarters of the UN Psychodeviant Police, there was no need to wear a throat mike, which had a tendency to make the voice sound muffled in spite of the Statistical Information-Bit Samplers which were supposed to clarify the speech coming through them.

"What do you know about 1968?" Houston asked sardonically. "Your mother was still pushing you around in a baby-carriage then."

"In a pram," corrected the Headquarters operator. "That is true, but my dear Aunt Jennifer told me all about it. She was—"

"The hell with your Aunt Jennifer," Houston interrupted suddenly. "Here comes Sir Lewis. Get me cover—fast!"

"Right. Keep us posted."

Sir Lewis Huntley stepped out of the broad door of the bank and turned left. He took a couple of steps and stopped. He didn't look around; he simply took a cigarette out of a silver case, put it in his mouth, and lit it. The glow of the lighter shone yellowly on the brass plate near the door which said: *An Affiliate of Westminster Bank, Ltd.*

Sir Lewis snapped the light out, drew on the cigarette, and strode on down the street, swinging a blue plastex brief case which contained a thousand pounds in United Nations Bank of England notes.

Houston decided the baronet had not been looking for a tail; he wished he could probe the man's mind to make sure, but he knew that would be fatal. He'd have to play the game and hope for the best.

"He's heading east," Houston whispered. "Doesn't look as if he's going to get a cab."

"Check," said the earphone.

Sir Lewis seemed in no great hurry, but he walked briskly, as though he had a definite destination in mind.

After a little way, he crossed to the south side of Leadenhall Street and kept going east. Houston stayed far enough behind to be above suspicion, but not so far that he ran a chance of losing his man.

"He's turning south on Fenchurch," Houston said a little later. "I wonder where he's going."

"Keep after him," said Headquarters. "Our net men haven't spotted either of you yet. They can hardly see across the street in this damned fog."

Houston kept going.

"What the hell?" he whispered a few minutes later. "He's still following Fenchurch Street! He's doubling back!"

Leadenhall Street, the banking center of the City of London, runs almost due east-and-west; Fenchurch Street makes a forty-five degree angle with it at the western end, running southwest for a bit and then curving toward the west, toward Lombard.

"Houston," said HQ, "touch your left ear."

Houston obediently reached up and scratched his left ear.

"Okay," said HQ. "Bogart's spotted you, but he hasn't spotted Sir Lewis. Bogart's across the street."

"He can't miss Sir Lewis," whispered Houston. "Conservatively dressed—matching coat and trousers of orange nylon tweed—royal blue half-brim bowler—carrying a blue brief case."

There was a pause, then: "Yeah. Bogart's spotted him, and so has MacGruder. Mac's on your side, a few yards ahead."

"Check. How about the rest of the net?"

"Coming, coming. Be patient, old man."

"I *am* patient," growled Houston. *I have to be,* he thought to himself, *otherwise I'd never stay alive.*

"We've got him bracketed now," HQ said. "If we lose him now, he's a magician."

Sir Lewis walked on, seemingly oblivious to the group of men who had surrounded him. He came to the end of Fenchurch Street and looked to his left, towards London Bridge. Then he glanced to his right.

"I think he's looking for a cab," Houston whispered.

"That's what MacGruder says," came the reply. "We've got Arthmore in a cab behind you; he'll pick you up. MacGruder will get another cab, and we have a private car for Bogart."

Sir Lewis flagged a cab, climbed in, and gave an address to the driver. Houston didn't hear it, but MacGruder, a heavy-set, short, balding man, was standing near enough to get the instructions Sir Lewis had given to the driver.

A cab pulled up to the curb near Houston, and he got in.

Arthmore, the driver, was a thin, tall, hawknosed individual who could have played Sherlock Holmes on TV. Once he got into character for a part, he never got out of it unless absolutely necessary. Right now, he was a Cockney cab-driver, and he would play the part to the hilt.

"Where to, guv'nor?" he asked innocently.

"Buckingham Palace," said Houston. "I've got a poker appointment with Prince Charles."

"Blimey, guv'nor," said Arthmore. "You *are* movin' in 'igh circles! 'Ow's 'Er Majesty these days?"

The turboelectric motor hummed, and the cab shot off into traffic. "According to the report I get on the blinkin' wireless," he continued, "a chap named MacGruder claims that the eminent Sir Lewis 'Untley is 'eaded for Number 37 Upper Berkeley Mews."

"One of these days," said Houston, "all those *H*'s you drop is going to bounce back and hit you in the face."

"Beg pardon, Mr. Yewston?" Arthmore asked blankly.

Houston grinned. "Nothing, cabbie; it's just that you remind me of a cultured, intelligent fellow named Jack Arthmore. The only difference is that Jack speaks the Queen's English."

"Crikey!" said Arthmore. "Wot a coincidence!" He paused, then: "The Queen's English, you say? She *'as* to be, don't she?"

"Shut up," said Houston conversationally. "And give me a cigarette," he added.

"There's a package of Players in my shirt pocket," Arthmore said, keeping his hands on the wheel.

Houston fished out a cigarette, lit it, and returned the pack.

Apropos of nothing, Arthmore said: "Reminds me of the time I was workin' for a printer, see? We 'ad to print up a bunch of 'andbills advertisin' a church charity bazaar. Down at the bottom was supposed to be printed 'Under the auspices of St. Bede's-on-Thames.' So I—"

He went on with a long, rambling tale about making a mistake in printing the handbill. Houston paid little attention. He smoked in silence, keeping his eyes on the red glow of the taillight ahead of them.

Neither man mentioned the approaching climax of the chase. Even hardened veterans of the Psychodeviant Police don't look forward to the possibility of having their minds taken over, controlled by some outside force.

It had never happened to Houston, but he knew that Arthmore had been through the experience once. It evidently wasn't pleasant.

"—and the boss was 'oppin' mad," Arthmore was saying, "but, crikey, 'ow was I to know that *auspice* was spelled A-U-S-P-I-C-E?"

Houston grinned. "Yeah, sure. How're we doing with Sir Lewis?"

"Seems to be headed in the right direction," Arthmore said, suddenly dropping the Cockney accent. "This is the route I'd take if I were headed for Upper Berkeley Mews. He probably hasn't told the driver to change addresses—maybe he won't."

"The victims never do," Houston said. "He probably is actually headed toward Number 37 Upper Berkeley Mews."

"Yeah. Nobody's perfect," said Arthmore.

Forty-five minutes of steady progress through the streets of Greater London brought Sir Lewis Huntley to Upper Berkeley and to the short dead-end street which constituted the Mews. By the time the dapper baronet stepped out of the machine and paid his driver, the whole area was surrounded by and filled with the well-armed, silent, and careful agents of the Psychodeviant Police.

Number 37 was an old concrete-and-steel structure of the George VI period, faced with a veneer of red brick. It had obviously been remodeled at least once to make the façade more modern and more fashionable; the red-violet anodized aluminum was relatively fresh and unstained. It wouldn't have taken vast wealth to rent a flat in the building, but neither would an average income have been quite enough.

Houston looked out of the window of Arthmore's cab and glanced at the tiers of windows in the building. Presumably, the man they were looking for was up there—somewhere.

So you occupy a station in the upper middle-class, thought Houston. It checked. Every bit of evidence that came his way seemed to check perfectly and fit neatly into the hypothesis which he had formed. Soon it would be time to test that theory—but the time had not yet come.

"Stand by and wait for orders, Houston," said the speaker in Houston's ear. "We've got men inside the building."

Sir Lewis Huntley opened the sparkling, translucent door of Number 37 Upper Berkeley Mews and went inside.

Arthmore pulled the cab over to the curb a few yards from the entrance and the two men waited in silence. All around them were other men, some in private cars, some walking slowly along the street. All of them were part of the net that had gathered to catch one man.

Poor fish, Houston thought wryly.

There was no noise, no excitement. Five minutes after Sir Lewis had entered the front door, it opened again. A man whom Houston had never seen before stepped out and gestured with one hand. At the same time, Houston's speaker said: "They've got him. Hit him with a stun gun when he tried to get out through the fire exit."

An ambulance which had been waiting at the entrance of the Mews pulled up in front of Number 37, and a minute or so later a little clot of men came out bearing a stretcher, which was loaded into the ambulance. Immediately after them came another man who had a firm, but polite grip on the arm of Sir Lewis Huntley.

Houston sighed and leaned back in his seat. That was that. It was all over. Simple. Nothing to it.

Another Controller had been apprehended by the Psychodeviant Police. Another deviant, already tried and found guilty, was ready to be exiled from Earth and imprisoned on one of the Penal Asteroids. All in the day's work.

There's just one thing I'd like to know, Houston thought blackly. *What in the hell's going on?*

In his hotel room near Piccadilly Circus, several hours later, David Houston sat alone, drink in hand, and put that same question to himself again.

"What's going on?"

On the face of it, it was simple. On the face of it, the answer was right in front of him, printed in black and white on the front page of the evening *Times.*

Houston lifted the paper off the bed and looked at it. The banner line said: *Controller Captured in Lambeth!*

Beneath that, in smaller type, the headline added: Robert Harris Accused of Taking Control of Barrister Sir Lewis Huntley.

The column itself told the whole story. Mr. Robert Harris, of No. 37 Upper Berkeley Mews, had, by means of mental control, taken over the mind of Sir Lewis and compelled him to draw one thousand pounds out of his bank. While Sir Lewis was returning to Harris with the money, the United Nations Psychodeviant Police had laid a trap. Sir Lewis, upon recovering his senses when Harris was rendered unconscious by a stun gun, had given evidence to the PD Police and to officials at New Scotland Yard.

Houston looked at the full-color photo of Harris that was printed alongside the column. Nice-looking chap; late twenties or early thirties, Houston guessed. Blond-red hair, blue eyes. All-in-all, a very pleasant, but ordinary sort of man.

There had been evidence that a Controller had been at work in London for some weeks now. Twelve days before, several men, following an impulse, had mailed twenty pounds to a "Richard Hempstead," General Delivery, Waterloo Station. By the time the matter had come to the authorities' attention, the envelopes had been called for and the Controller had escaped.

Robert Harris was not the first Controller to be captured, nor, Houston knew, would he be the last. The first one had shown up more than sixteen years before, in Dallas, Texas, USA.

Houston grinned as he thought of it. Projective telepathy had only been a crackpot's idea back then. In spite of the work of many intelligent, sane men, who had shown that mental powers above and beyond the ordinary did exist, the average man simply laughed off such nonsense. It was mysticism; it was magic; it was foolish superstition. It was anything but true.

But ever since "Blackjack" Donnely had practically taken control of the whole city of Dallas, the average man had changed his mind. It was still mysterious; it was still magic; but now the weird machinations of the supernormal mind were something to be feared.

In the sixteen years that had ensued since the discovery of the abnormal mental powers of "Blackjack" Donnely, rumors had spread all over the world. There were supposed to be men who could levitate—fly through the air at will. Others could walk through walls, and still others could make themselves invisible. The horrible monsters that were supposed to be walking the Earth were legion.

Actually, only one type of supernormal psychodeviant had been found—the telepath, the mindreader who could probe into the mental processes of others. Worse than that, the telepath could project his own thoughts into the mind of another, so that the victim supposed that the thoughts were his own. Actually, it was a high-powered form of hypnotism; the victim could be made to do anything the projective telepath wanted him to.

"Blackjack" Donnely had made that clear in his trial in Texas.

Donnely had been a big man—big physically, and important in city politics. He had also been as arrogant as the Devil himself.

It was the arrogance that had finally tripped up Donnely. He had thought himself impregnable. Haled into court on charges of misappropriation of public funds, he had just sat and smirked while several witnesses for the State admitted that they had aided Donnely, but they claimed he had "hypnotized" them. Donnely didn't try to interfere with the evidence—that's where he made his mistake. And that's where his arrogance tripped him up.

If he'd used telepathic projection to influence the State Attorney or the witnesses or the judge or the Grand Jury *before* the trial, he might never have been discovered as the first of the Controllers. But that wasn't Donnely's style.

"None of this namby-pamby stuff," he had once been quoted as saying; "if you got enemies, don't tease 'em—show 'em who's running things. Blackjack 'em, if you have to."

And that's exactly what "Blackjack" Donnely had done. The trial was a farce from beginning to end; each witness gave his evidence from the stand, and then Donnely took control of their minds and made them refute every bit of it, publicly and tearfully apologizing to the "wonderful Mr. Donnely" for saying such unkind things about him.

The judge and the jury knew something funny was going on, but they had no evidence, one way or another. The case, even at that point, might have ended with an acquittal or a hung jury, but Donnely wasn't through using his blackjack.

He took over the mind of the foreman of the jury. The foreman claimed later that the jury had decided that they could reach no decision. Other jurors claimed that they had decided Donnely was guilty, but that was probably an *ex post facto* switch. It didn't matter, anyway; when the foreman came out, he pronounced Donnely innocent. That should have ended it.

The other jurors began to protest, but by that time, Donnely had gained control of the judge's mind. Rapidly, the judge silenced the jurors, declared Donnely to be free, and then publicly apologized for ever daring to doubt Mr. Donnely.

The State's Attorney was equally verbose in his apology; he was almost in tears because of his "deep contrition at having cast aspersions on the spotless character of so great a man."

Donnely was released.

The next evening, "Blackjack" Donnely was shot down at the front door of his own home. There were fifteen bullets in his body; three from a .32, five from a .38, and seven from a .45.

The police investigation was far from thorough; any evidence that may have turned up somehow got lost. It was labelled as "homicide committed by person or persons unknown," and it stayed that way.

Donnely was only the first. In the next two years, four more showed up. Everyone of them, in one way or another, had attempted to gain power or money by mental projection. Everyone of them was a twisted megalomaniac.

Houston looked again at Harris's picture on the front page of the *Times*. Here was one Controller who neither looked nor acted like a megalomaniac. That wouldn't make much difference to the PD Police; as far as the officials were concerned, the ability to project telepathically and the taint of delusions of grandeur went hand in hand. Controllers were power-mad and criminal by definition.

Fear still ruled the emotional reactions against Controllers, in spite of the protection of the Psychodeviant Police.

But David Houston knew damned good and well that all telepaths were not necessarily insane.

He should know. He was a Controller, himself.

Brrrring!

David Houston tossed the paper on the bed and walked over to the phone. He cut in the circuit, and waited for the phone's TV screen to show the face of his caller. But the screen remained blank.

"Who is it?" Houston asked.

"Is this CHAring Cross 7-8161?" It was a woman's voice, soft and well-modulated.

"No, this is CHElsea 7-8161," Houston said. "You must have dialed C-H-E instead of C-H-A."

"Oh. I'm very sorry. Excuse me." There was a click, and she hung up.

Houston walked back over to the bed and picked up his paper. He looked at it, but he didn't read it. It no longer interested him.

So Dorrine was finally in London, eh? He'd recognized her voice instantly; even years of training couldn't smother the midwestern American of Chicago completely beneath the precise British of the well-educated English girl.

The signal had been agreed upon, just in case his phone was tapped. Even the Psychodeviant Police could be suspected of harboring a Controller—although Houston didn't think it too likely. Nevertheless, he wasn't one to take too many chances.

He glanced at his watch. He had an hour yet. He'd wait five minutes before he phoned headquarters.

He sat down in his chair again and forced himself to relax, smoke a cigarette, and read the paper—the sports section. Perusing the records of the season's cricket matches kept his mind off that picture on the front page. At least, he hoped they would. Let's see, now—Benton was being rated as the finest googly bowler on the Staffordshire Club ...

Everything went fine until he came across a reference to a John Harris, a top-flight batsman for Hambledon; that reminded him of Robert Harris. Houston threw down the paper in disgust and walked over to the phone.

The number was TROwbridge 5-4321, but no one ever bothered to remember it. Simply dial 8-7-6-5-4-3-2-1, and every time a voice at the other end would answer—

"Hamilton speaking."

"Houston here; will I be needed in the next hour or so?"

"Mmmm. Just a second; I'll check the roster. No; your evidence won't be needed personally. You've filed an affidavit. No, I don't think—wait a minute! Yes, there's a return here for you; reservation on the six A.M. jet to New York. Your job here is done, Houston, so you can take the rest of the evening off and relax. Going anywhere in particular?"

"I thought I'd get a bite to eat and take in a movie, maybe, but if I'm due out at six, I'll forego the cinematic diversion. When's the trial?"

"It's scheduled for eleven-thirty this evening. Going to come?"

Houston shook his head. "Not if I'm not needed to give evidence. Those Controllers always give me the creeps."

"They do everybody," said Hamilton. "Well, you caught him; there's no need for you to stick around for the windup. Have a good time."

"Thanks," said Houston shortly, and hung up.

The windup, Houston thought. *Sure. That's all it will be. A Controller's trial is a farce. Knock him out with a stun gun and then pump him full of comatol. How can he defend himself if he's unconscious all through the trial?*

Houston knew what the average man's answer to that would be: "If a Controller were allowed to remain conscious, he'd take over the judge's mind and get himself freed."

Houston said an obscene word under his breath, jammed his hat on his head, put on his coat, and left his apartment.

With the coming of darkness, the heavy fog had become still denser. The yellow beams of the sodium vapor lamps were simply golden spots hanging in an all-enveloping blackness. Walking the street was a process of moving from one little golden island of light to another, crossing seas of blankness between. The monochromatic yellow shone on the human faces that passed beneath the lamps, robbing them of all color, giving them a dead, grayish appearance beneath the yellow itself.

David Houston walked purposefully along the pavement, his hand jammed deep in his overcoat pockets. One hand held the control box for the little earpiece he wore. He kept moving the band selector, listening for any sign that the Psychodeviant Police were suspicious of a Controller in their midst.

If they were following him, of course, they would use a different scrambler circuit than the one which was plugged into his own unit, but he would be able to hear the gabble of voices, even if he couldn't understand what they were saying.

So far, there hadn't been a sound; if he was being followed, his tailers weren't using the personal intercom units.

He didn't try to elude anyone who might be following. That, in itself, would be a giveaway. Let them watch, if they *were* watching. They wouldn't see anything but a man going to get himself a bit of dinner.

The *Charles II Inn*, on Regent Street, near Piccadilly Circus, was a haven of brightness in an otherwise Stygian London. It was one of those "old-fashioned" places—Restoration style of decoration, carried out in modern plastics. The oak panelling looked authentic enough, but it was just a little too glossy to be real.

Houston pushed open the door, stepped inside, removed his hat and coat and shook the dampness from them. As he handed them to the checker, he looked casually around. Dorrine was nowhere in sight, but he hadn't expected her to be. There would be no point in their meeting physically; it might even be downright dangerous.

The headwaiter, clad in the long waistcoat and full trunk-hose of the late Seventeenth Century, bowed punctiliously.

"You're alone, sir?"

"Alone, yes," Houston said. "I'll just be wanting a light supper and a drink or two."

"This way, sir."

Houston followed the man to a small table in the rear of the huge dining room. It was set for two, but the other place was quickly cleared away. Houston ordered an Irish-and-soda from a waiter who was only slightly less elaborately dressed than the headwaiter, and then settled himself down to wait. If he knew Dorrine, she would be on time to the minute.

She came while the waiter was setting the drink on Houston's table. She stepped in through the door, her unmistakable hair glowing a rich red in the illumination of the pseudo-candlelight.

She didn't bother to look around; she knew he would be there.

After a single glance, Houston averted his eyes from her and looked back at his drink.

And in that same instant, their minds touched.

Dave, darling! I knew you'd be early!

Dorrine!

And then their minds meshed for an instant.

I—(we)—you—LOVE—*you*—(each other)—me!—us!

Houston looked complacently at his drink while the headwaiter led Dorrine to a table on the far side of the room. She sat down gracefully, smiled at the waiter, and ordered a cocktail. Then she took a magazine from her handbag and began—presumably—to read.

Her thought came: *Who is this Richard Harris? He's not one of our Group.*

Houston sipped at his drink. *No. An unknown, like the others. I wonder if he's even a telepath.*

What? Her thought carried astonishment. *Why, Dave—he'd* have *to be! How else could he have controlled this Sir Lewis—whatsisname—Huntley?*

Well—I've got a funny idea, Houston replied. *Look at it this way: So far as we know, there are two Groups of telepaths. There's our own Group. All we want is to be left alone. We don't read a Normal's mind unless we have to, and we don't try to control one unless our lives are threatened. We stay under cover, out of everyone's way.*

Then there are the megalomaniacs. They try, presumably, to gain wealth and power by controlling Normals. And they get caught with monotonous regularity. Right?

The girl caught an odd note in that thought. *What do you mean, "monotonous regularity"?* she asked.

I mean, Houston thought savagely, *why is it they're all so bloody stupid? Look at this Harris guy; he is supposed to have taken over Sir Lewis's mind in order to get a thousand pounds. So what did he have Sir Lewis do? Parade all around the city to pick up a PD Police net, and then give his address to a cabman in a loud voice and lead the whole net right to Harris! How stupid can a man get?*

It does look pretty silly, Dorrine agreed. *Have you got an explanation?*

Several, Houston told her. *And I don't know which one is correct.*

Let's have them, the girl thought.

Houston gave them to her. None of them, he knew, was completely satisfactory, but they all made more sense that the theory that Harris had done what the PD Police claimed he'd done.

Theory Number One: The real megalomaniac Controller had taken over Sir Lewis's mind and made him draw out the thousand pounds and head west on Leadenhall Street. Somehow, the Controller had found out that Sir Lewis was being followed, and had steered him away from the original destination, heading him toward the innocent Robert Harris. That implied that the Controller had been within a few dozen yards of the net men that afternoon. A Controller can't control a mind directly from a distance, although orders can be implanted which will cause a man to carry out a plan of action, even though he may be miles from the Controller. But in order to change those plans, the Controller would have to be within projection range.

Theory Two: Robert Harris actually was a megalomaniac Controller; with a long record of success behind him, who had finally grown careless.

At that point, Dorrine interjected a thought: *Isn't it possible that he wanted to be caught?*

Houston mulled it over for a minute. *A guilt-punishment reaction? He wanted to be punished for his crimes? I suppose that might account for part of it, yes. But if he'd been so successful, what did he do with all his money?*

Dorrine gave a mental shrug. *Who knows? What's Theory Number Three?*

Number Three was the screwiest one of all, yet it made a weird kind of sense. Suppose that Sir Lewis himself had had a grudge against Harris? The whole thing would have been ridiculously easy; all he'd have to do would be to act just as he had acted and then give evidence against Harris.

The thing that made it odd wasn't the actual frame-up (if that's what it was); these days, every crime was blamed on a Controller. A man accused of murder simply looked virtuous and said that he would never have done such a thing if he hadn't been under the power of a Controller. Ditto for robbery, rape, and any other felony you'd care to name.

An aura of fear hung over the whole Earth; each man half suspected everyone with whom he came in contact of being a Controller.

So it wasn't that the frame-up in itself was peculiar in this case; it was simply that it wasn't Sir Lewis Huntley's style. If Sir Lewis had wanted to get Harris, he'd have done it legally, without any underhanded frame-ups. Still, the theory remained as a possibility.

I suppose it does, Dorrine agreed, *but how does that tie in with our own Group? What about Jackson and Marcy? What happened to them?*

I don't know, Houston admitted, *I just don't know.*

Jackson and Marcy had been members of the Group of telepaths who had banded together for companionship and mutual protection. Both of them had been trapped by the PD Police in exactly the same way that Harris had been trapped. They were now where Harris would be in a matter of hours—in the Penal Cluster.

Their arrests didn't make sense, either; they had been accused of taking over someone's mind for the purpose of gaining money illegally—illegal, that is, according to the new UN laws that had been passed to supersede the various national laws that had previously been in effect.

But Houston had known both men well, and neither of them was the kind of man who would pull such a stunt, much less do it in such a stupid manner.

Dorrine thought: *Well, Dave, this Harris case is out of our hands now; we've got to concentrate on getting others into the Group—we've got to find the other sane ones.*

You're ready to take over here, then? he asked.

At the table, several yards away from where Houston was sitting, Dorrine, still looking at the book, smiled faintly.

I'll have to; you're being transferred back to New York at six in the morning.

Houston allowed a feeling of startled surprise to bridge the gap between their minds. *How'd you know that?* He hadn't told her, and she couldn't have forced the knowledge from his mind. A telepath can open the mind of a Normal as simply as he might open the pages of a book, but the mind of another Controller is far stronger. One telepath couldn't force anything from the mind of another; all thoughts had to be exchanged voluntarily.

She was still smiling. *We've got a few spies in the UN now,* she told him. *I got the information before you did.*

You knew before you left New York? he asked incredulously.

That's right, she thought. *The decision was made last night. Why?*

Nothing, he told her. *I was just surprised, that's all.* But deep behind the telepathic barrier he had erected against her probing mind, he was thinking something else. He had been assigned to London to capture the Controller—then unknown—who was said to be active in England. But his recall order had been decided upon *before* Harris was caught—or even suspected. Someone in the UN Psychodeviant Police Supreme Headquarters in New York must have known that Harris would be caught that day!

Something's bothering you, Dorrine stated flatly.

I was thinking about leaving London, he replied evasively. *I haven't seen you for six months, and now I have to leave again.*

I'll be back in New York within three weeks, the girl thought warmly. *I'll be—*

Her thoughts were cut off suddenly by a strident voice in Houston's ear. "Attention; all-band notice. Robert Bentley Harris, arraigned this evening on a charge of illegal use of psychodeviant powers for the purpose of compounding a felony, has been found guilty as charged. He was therefore sentenced by the Lord Justice of Her Majesty's Court of Star Chamber to be banished from Earth forever, such banishment to be carried out by the United Nations Penology Service at the Queen's pleasure."

The words that were running through Houston's brain, had been transmitted easily to Dorrine. For a moment, neither of them made any comment. Then Houston glanced at his watch.

Twenty-one minutes, he thought bitterly. *What took them so long?*

High in the thin ionosphere, seventy miles above the surface of the Earth, a fifteen-hundred-mile-an-hour rocket airliner winged its way westward across the Atlantic, pushing herself forward on the thin, whispering, white-

hot jeíarris was a Controller, all right; there was no doubt about that. But below e in his mind was there any trace of any knowledge of what had

Da ed to Sir Lewis Huntley. If Sir Lewis had actually been controlled, it quietly i een done by Robert Harris.

When Ha ton wished he'd been able to probe Sir Lewis's mind; he'd have New Yorl o get a lot more information out of it than he had in his possession to send fo hat would have been dangerous; if Sir Lewis was a Controller

But nc d had been acting a part, Houston would have given himself hours befonstant he attempted to touch the baronet's mind. If, on the other

Did so wis had actually been under the control of another telepath, any captured? C the mind of the puppet would have betrayed Houston to the about Pol er.

ew nothing. He wasn't acquainted with any other Controllers, is nose clean ever since he'd discovered his latent powers. He alomaniac Controllers were either captured or mobbed, and to experience either.

Normals had long since discovered that the only way to overcome Controller was by force of numbers. A Controller could only hold one Normal mind at a time. That was why a mob could easily kill a single Controller; that was why the Psychodeviant Police had evolved the "net" system for arresting a telepath.

Harris, then, had been framed. Or could it be called a frame-up when Harris was really guilty of the actual crime? Because the crime he had really been accused of was not that of controlling Sir Lewis, but the crime of being a telepath. That, and that alone, damned him in the eyes of the Normals; the crime of taking over a mind for gain was incidental. The stigma lies in what he *was*, not what he did.

Harris himself was in the bottom of the plane, in the baggage section near the landing gear. After his trial, still drugged, he had been secretly put aboard, to be taken to the Long Island Spaceport in New York. It had had to be secret; no Normal would knowingly ride on an aircraft which carried a Controller, even if he were drugged into total unconsciousness.

With Harris were two PD Police guards. Their low conversation impinged on Harris's ears, and was transmitted to Houston's mind.

Suddenly, one of them said: "Hey! He's moving!"

"Better give him another shot, Harry;" said the other, "when those guys wake up, they drive you crazy."

Houston could almost feel the sting of the needle as it was inserted into the arm of the helpless prisoner.

Slowly, Harris's thoughts, which had begun to become fully coherent, again became chaotic, finally sliding off into silence and darkness.

"Are you all right, sir?"

Houston looked up from his intense concentration. The stewardess was standing by his seat. He realized that there was a film of perspiration on his brow, and that he probably had looked dazed while he was concentrating on Harris's mind.

"Sure," he said quickly, "I'm all right. I'm just a little tired. Had to get up too early to catch this plane." He rubbed his forehead. "I do have a little headache; would you happen to have any aspirin aboard?"

She smiled professionally. "Certainly, sir. I'll get a couple of tablets."

As she left for the first-aid cabinet, Houston thought bleakly to himself: *Harris was framed. Possibly others have been, too. But by whom? And why?*

He could see why a Normal might do such a thing. But why would a Controller do it?

There was only one answer. Somewhere, there was a Controller, or a group of Controllers who were megalomaniacs *par excellence*. If that were so, he—or they—could make the late "Blackjack" Donnely look like a meek, harmless, little mouse.

The one part of Continental U.S.A. over which the American Government had no jurisdiction was small, areawise, in comparison with its power. The District of the United Nations occupied the small area of Manhattan Island which ran from 38th Street on the south to 49th Street on the north; its western border was Third Avenue, its eastern, the East River. From here, the UN ruled Earth.

There were no walls or fences around it; only by looking at street signs could anyone tell that they had crossed an international border. Crossing Third Avenue from west to east, one found that 45th Street had suddenly become Deutschland Strasse; 40th Street became Rue de France; 47th was the Via Italiano. 43rd Street's sign was painted in Cyrillic characters, but beneath it, in English, were the words "Avenue of Mother Russia."

Third Avenue was technically One World Drive. Second Avenue was labelled as Planetary Peace Drive, and First was United Nations Drive.

But New Yorkers are, and always have been, diehards. Just as The Avenue of the Americas had forever remained Sixth Avenue, no matter what the maps called it, so had the other streets retained their old names in conversation.

Even the International Post Office, after years of wrangling, had given up, and letters addressed to *Supreme Headquarters, United Nations Police, 45th Street at Second Avenue,* were delivered without comment, even though the IPO still firmly held that they were technically misaddressed. And, privately, even the IPO officials admitted that the numbers were easier to say and remember than the polyglot street names that had been tagged on by the General Assembly.

So when David Houston signalled a taxi at Grand Central Station and said, "Forty-fifth and Second," the driver simply set his automatic controls, leaned back in his seat, and said, "Goin' to see the cops, huh?"

When no answer was forthcoming, the driver turned around and took a good look at his passenger. "Maybe you're a UN cop yourself, huh?"

Houston shook his head. "Nope. Some kids have been scribbling dirty words on my sidewalk, and I'm going to report it to the authorities."

The driver turned back around and looked ahead again. "Jeez! That's serious. Hadn't you better take it up with the Secretary General? I wouldn't be satisfied with no underlings in a case like that."

"I'm thinking of taking it up with the Atomic Energy Control Board," Houston told him. "I think those kids are using radioactive chalk."

"That's one way for 'em to get blue jeans," said the driver cryptically.

There was silence for a moment as the taxi braked smoothly to a halt, guided and controlled by the automatic machinery in the hood.

Then, suddenly, the driver said: "Ship up!" He pointed east, along 45th Street, toward Long Island. Far in the distance was a rapidly rising vapor trail, pointing vertically toward the sky, the unmistakable sign of a spaceship takeoff. They didn't leave often, and it was still an unusual sight.

Houston said nothing as he climbed out and paid the driver, tossing in an extra tip.

"Thanks, buddy," said the driver. "Watch out for them kids."

Houston didn't answer. He was still watching the vapor trail as the cab pulled away.

There goes Harris, he was thinking. *An innocent man, guilty of nothing more than being born different. And because of that, he's labelled as an inhuman monster, not even worthy of being executed. Instead, he's taken into space, filled full of hibernene, and chained to a floating piece of rock for the rest of his life.*

Such was humanity's "humane" way of taking care of the bogey of Controllers. Capital punishment had been outlawed all over Earth; it had long since been proved that legalized murder, execution by the State, solved nothing, helped no one, prevented no crimes, and did infinitely more harm than good in the long run.

With the coming of the Controllers, a movement had arisen to bring back the old evil of judicial murder, but it had been quickly put down when the Penal Cluster plan had been put forth as a more "humane" method.

Hibernene was a drug that had been evolved from the study of animals like the bear, which spent its winters in an almost death-like sleep. A human being, given a proper dosage of the drug, lapsed into a deep coma. The bodily processes were slowed down; the heart throbbed sluggishly, once every few minutes; thought ceased. It was the ideal prison for a mental offender that ordinary prisons could not hold.

But it wasn't quite enough for the bloodthirsty desire for vengeance that the Normals held for the Controllers. There had to be more.

Following Earth in its orbit around the sun, trailing it by some ninety-three million miles, were a group of tiny asteroids, occupying what is known as the Trojan position. They were invisible from Earth, being made of dark rock and none of them being more than fifteen feet in diameter. But they had been a source of trouble in some of the early expeditions to Mars, and had been carefully charted by the Space Commission.

Now a use had been found for them. A man in a spacesuit could easily be chained to one of them. With him was a small, sun-powered engine and tanks of liquified food concentrates and oxygen. Kept under the influence of hibernene, and kept cool by the chill of space, a man could spend the rest of his life there—unmoving, unknowing, uncaring, dead as far as he and the rest of Mankind were concerned—his slight bodily needs tended automatically by machine.

It was a punishment that satisfied both sides of the life-or-death argument.

Houston shook off the bleak, black feeling of terrible chill that had crept over him and pushed his way into the UN Police building.

The thirteenth floor housed the Psychodeviant Division. As he stood in the rising elevator, Houston wondered wryly if the number 13 was good luck or bad in this case.

He stepped out of the elevator and headed for the Division Chief's office.

Division Chief Reinhardt was a heavy-set, balding man, built like a professional wrestler. His cold blue eyes gleamed from beneath shaggy, overhanging brows, and his face was almost expressionless except for a faint scowl that crossed it from time to time. In spite of the fact that a Canadian education had wiped out all but the barest trace of German accent, his Prussian training, of the old Junkers school, was still evident. He demanded—and got—precision and obedience from his subordinates, although he had no use for the strictly military viewpoint of obsequiousness towards one's superiors.

He was sitting behind his desk, scowling slightly at some papers on it when Houston stepped in.

"You wanted me to report straight to you, Mr. Reinhardt?"

Reinhardt looked up, his heavy face becoming expressionless. "Ah, Houston. Yes; sit down. You did a fine job on that London affair; that's what I call coming through at the last moment."

"How so?"

"Your orders to return," he said, "were cut before you found your man. We have a much more important case for you than some petty pilfering Controller. We are after much more dangerous game."

Houston nodded. "I see." Inwardly, he wondered. It was almost as if Reinhardt knew that Houston had found out that the recall had come early. Houston would have given his right arm at that moment to be able to probe Reinhardt's mind. But he held himself back. He had, in the past, sent tentative probes toward the Division Chief and found nothing, but he didn't know whether it would be safe now or not. It would be better to wait.

Reinhardt stood up, walked to the wall, and turned on a display screen. He twisted a knob to a certain setting, and a map of Manhattan Island sprang onto the screen in glowing color.

"As you know," Reinhardt said pedantically, "no Controller can do a perfect job of controlling a normal person. No matter how much he may want to make John Smith act naturally, some of the personality of the Controller will show up in the actions of John Smith. Am I correct?"

Houston nodded without saying anything. The question was purely rhetorical, and the statement was perfectly correct.

"Very well, then," Reinhardt continued, "by means of these peculiarities, our psychologists have found that there is widespread, but very subtle controlling going on right in the UN General Assembly itself! The amazing thing is that they all bear the—shall we say—trademark of the same Controller. Whoever he is, he seems to have a long-range plan in mind; he wants to change, ever so slightly, certain international laws so that he will profit by them. Do you follow?"

"I follow," said Houston.

"Good. It has taken painstaking research and a great deal of psychological statistical analysis, but we have found that one company—and one company only—benefits by these legal changes. Did you ever hear of Lasser & Sons?"

"Sure," said Houston. "They're in the import-export business, with a few fingers in shipping and air transport."

"That's them," said Reinhardt. "Someone in that company, presumably someone at the top, is a Controller. And he's a very subtle, very dangerous man. Unlike the others, there is nothing hasty or overt in his plans. But within a few years, if this goes on, he will have more power than the others ever dreamed of."

"And my job is to get him?" Houston asked.

Reinhardt nodded. "That's it. Get him. One way or another. You're in charge; I don't care how you do it, but this one Controller is more dangerous than any other we've come across, so get him."

Houston nodded slowly. "Okay. Can you give me all the data you have so far?"

Reinhardt patted a heavy folder on his desk. "It's all here." Then he tapped the projected map on the screen. "That's the Lasser Building—Church Street at Worth. Somewhere in there is the man we're looking for."

David Houston spent the next six weeks gathering facts, trying to determine the identity of the mysterious Controller at Lasser & Sons. Slowly, the evidence began to pile up.

At the same time, he worried over his own problem. Who was betraying non-criminal Controllers to the PD Police?

In that six-week period, two more men and a woman were arrested—one in Spain, one in India, and one in Hawaii.

There weren't very many Controllers on Earth, percentagewise. Of the three and a half billion people on Earth, less than an estimated one-thousandth of one percent were telepathic. But that made a grand total of some thirty-five thousand people.

Spread, as they were, all over the planet, it was rare that one Controller ever met another. The intelligent ones didn't use their power; they remained concealed, even from each other.

But *someone*, somewhere, was finding them and betraying them to the Psychodeviant Police.

As more and more data came in on the Lasser case, Houston began to get an idea. If there were a really clever, highly intelligent, megalomaniac Controller, wouldn't it be part of his psychological pattern to attempt to get rid of the majority of Controllers, those who simply wanted to lead normal lives?

And, if so, wasn't it possible that both his cases—the official and the unofficial—might lead to the same place: Lasser & Sons?

It began to look as though Houston could kill both his birds at once, if he could just figure out when, how, and in what direction to throw the stone.

In the middle of the seventh week, a Controller in Manchester, England, was mobbed and torn to bits by an irate crowd before the PD Police could get to him. There was no doubt in Houston's mind that this one was a real megalomaniac; he had taken over another man's brain and forced him to commit suicide. The controlled man had taken a Webley automatic, put it to his temple, and blown his brains out.

The Controller's mistake was in not realizing what the sudden shock of that bullet, transmitted to him telepathically, could do to his own mind. In the mental disorder that followed, he was spotted and killed easily.

There was still no word from Dorrine. She had flown back to the States a week after Houston had returned, but she had had to get back to England after three days. Since then, he had had three letters, nothing more. And letters are a damned unsatisfactory way for a telepath to conduct a love affair.

The one other factor that entered in was The Group, the small band of sane, reasonable telepaths who had begun to build themselves into an organization—a sort of Mutual Protective Association.

Personally, Houston didn't think much of the idea; the Group didn't have any real organization, and they refused to put one together. It was supposed to be democratic, but it sometimes bordered on the anarchic.

He stayed with them more for companionship than any other reason. When Dorrine had come back for her short stay, Houston had met with them and tried to get them to help him trace down the megalomaniac

Controller who was doing so much damage, but they'd balked at the idea. Their job, they claimed, was to get enough members so that they could protect themselves from arrest by the Normals, and then just let things ride.

"After all," Dorrine had said, "things will work themselves out, darling; they always do."

"Not unless somebody helps them, they don't," Houston had snapped back. "Someone has to do something."

"But, Dave, darling—we *are* doing something! Don't you see?"

He didn't, but there was no convincing either the Group or Dorrine. She was passionately interested in the recruiting work she was doing, and she thought that the Group was the answer to every Controller's troubles.

And then she had rushed back to England. "I'll be back soon, Dave," she'd said. "I think I have a lead on a girl in Liverpool."

So far, the girl hadn't been found. Controllers didn't like to give themselves away to anyone, so they kept a tight screen up most of the time.

It seemed as though everyone on Earth was in deadly fear all the time. The Normals feared losing their identities to Controllers, and the Controllers feared death at the hands of the Normals.

And death or the Penal Cluster were their only choices if they were discovered.

Houston worried about the risks Dorrine was taking, but there was nothing he could do. She was doing what she thought was right, just as he was; how could he argue with that?

Houston went on with his job, putting together facts and rumors and statistical data analysis, searching out his quarry.

And, at the end of the eighth week, everything blew high, wide, and hellish.

It was late evening. A cool wind blew over New York, bringing with it a hint of the rain to come. Church Street, in lower Manhattan, was not crowded, as it had been in the late afternoon, but neither was it entirely deserted. The cafes and bars did a lively business, but the tall, many-colored office buildings gaped at the street with blind and darkened eyes. Only a few of the windows glowed whitely with fluorescent illumination.

In one of the small coffee shops, David Houston sat, smoking a cigarette and stirring idly at a cup of cooling coffee.

Across the street was the Lasser Building; high up on the sixtieth floor, a whole suite of offices was brightly lit. The rest of the building was clothed in blackness.

Who was up there in that suite? Houston wasn't quite sure. He had narrowed his list of suspects down to three men: John Sager, Loris Pederson, and Norcross Lasser, three top officials in the company. Sager and Pederson were both vice-presidents of the firm; Sager was in charge of the Foreign Exports department, while Pederson handled the actual shipping. Lasser, by virtue of being the grandson of the man who had founded the firm, was president of Lasser & Sons, Inc.

Lasser seemed like a poor choice as chief villain of the outfit; he was a mild, bland man, quiet and friendly. Besides, his position made him an obvious suspect; naturally, the majority stockholder of the firm would profit most by the increased power of the company. And, equally obviously, a Controller wouldn't want to put himself in such an exposed position.

Which made Lasser, in Houston's mind, a hell of a good suspect. If anything happened, Lasser could cover by claiming that he, too, had been controlled, and the chances were that he could get away with it. A Controller never did anything directly; their dirty work was done by someone else—a puppet under their mental control. At least, so ran the popular misconception. If Lasser were the man, he stood a good chance of getting away with it, even if he were caught, provided he played his cards right.

That reasoning still didn't eliminate Sager or Pederson. Either of them could be the Controller. And there still remained the possibility that some unknown, unsuspected fourth person had the company of Lasser & Sons under his thumb.

That was what Houston intended to find out tonight.

He took a sip of his coffee, found it still reasonably hot.

Damn the megalomaniacs, anyway! Houston subconsciously tightened his fists. He, personally, had more to fear from the Normals than from another Controller. Normals could kill or imprison him, while a Controller would have a hard time doing either, directly.

But Houston could understand the Normal man; he could see how fear of a Controller could drive a man without the ability into a frenzied panic. He could understand, even forgive their actions, born and bred in ignorance and fear.

No, the ones he hated were the ones who had conceived and fostered that fear—the psychologically unstable megalomaniac Controllers. There were only a handful of them—probably not more than a few hundred or a thousand. But because of them, every telepath on Earth found his life in danger, and every Normal found his life a hell of terror.

Let Bonnie and her do nothing friends run around the globe recruiting members for their precious Group; that was all right for them. Meanwhile, David Houston would be doing something on a more basic action level.

He glanced at his watch. Almost time.

"How's the deployment?" he whispered in his throat.

"We've got the building surrounded now," said the voice in his ear. "You can go in anytime."

"How about the roof?"

"That's taken care of, sir; we've got 'copter that can be on the top of the Lasser Building at any time you call. They can land within thirty seconds of your signal."

"Okay," Houston said; "I'm going in now. Remember—no matter what I say or do, no one is to leave that building if they're conscious. And keep your eyes on me; if I act in the least peculiar, handcuff me—but don't knock me out.

"And if I'm not back on time, come in anyway."

"Right."

Houston finished his coffee, dropped a coin on the counter, and headed for the other side of the street.

The big problem was getting into the building itself. It was ringed with alarms; Lasser & Sons didn't want just anybody wandering in and out of their building.

So Houston had arranged a roundabout way. The building next to the Lasser Building was a good deal smaller, only forty-five stories high. A week before, Houston had rented an office on the eighteenth floor of the building; on the door, he had already had a sign engraved: Ajax Enterprises.

It was a shame the office would never be used.

Houston walked straight to the next-door building and opened the front door with his key. Inside, a night watchman lounged behind a desk, smoking a blackened briar. He looked up, smiled, and nodded.

"Evening, Mr. Griswold; working late tonight?"

Houston forced a smile he did not feel. "Just doing a little paper work," he said.

He took the automatic elevator to the eighteenth floor. He didn't relish the idea of walking up to the roof, but taking the elevator would make the nightwatchman suspicious.

He didn't bother going to the office; he headed directly for the stairway and began his long climb—twenty-seven floors to the roof.

All through it, he kept up a running comment through his throat mike. "I wish I weighed about fifty pounds less; carrying two hundred and twenty pounds of blubber up these stairs isn't easy."

"Blubber, hooey!" the earphone interrupted. "Any man who's six-feet-three has a right to carry that much weight. Actually, you're a skinny-looking sort of goop."

Both men were exaggerating; Houston wasn't fat, but his broad, powerful frame couldn't be called skinny, either.

When he finally reached the roof, he paused and surveyed the wall of the Lasser Building, which towered high above him, spearing an additional thirty stories in the air. Up there, the lights on the sixtieth floor gleamed in the night.

The air was growing cooler, and the beginnings of a mist were forming. Houston hoped it wouldn't start to rain before he got inside.

The forty-sixth floor of the Lasser Building had no windows on this side, but there were plenty on the forty-seventh.

Leading up to them was an inviting looking fire escape, but Houston knew he didn't dare take that. By law, every fire escape was rigged with a fire alarm, in addition to the regular burglar alarm. He'd have to use another way.

The Lasser Building was a steel structure, shelled over with a bright blue anodized aluminum sheath. Only the day before, Houston, wearing the gray coverall of a power-line workman, had checked the wall to find the big steel beams beneath the aluminum. He had also installed certain other equipment; now he was going to make use of it.

Concealed in the louvres of the air-conditioner intake of the lower building was a specially constructed suit and several hundred feet of power line which was connected to the main line of the building.

In the darkness, Houston slipped on the suit. It was constructed somewhat like a light diving suit or a spacesuit, but without the helmet.

In the toes, knees, and hands, were powerful electromagnets controlled by switches in the fingers of the gloves and powered by the current in the long line.

Houston stepped over to the blue aluminum wall, reached out a hand, and lowered one finger. Instantly, the powerful magnet anchored his hand to the wall, held by the dense magnetic field to the steel beam beneath the aluminum sheath. That one magnet alone could support his full body weight, and he had six magnets to work with.

Slowly, carefully, David Houston began to crawl up the wall.

Turn on a magnet in the right hand; lift up the left hand and anchor it higher; turn on the right hand and lift it even with the left, then anchor it again; do the same with both legs; then begin the process all over again, turning the magnets off and on in rotation.

Up and up he went. Past the forty-sixth floor, past the forty-seventh, the forty-eighth, and the forty-ninth. Not until he reached the fiftieth floor did he attempt to open one of the windows.

There was a magnetic lock inside the window, but Houston had taken that, too, into account. The powerful magnet in his right glove slid it aside easily. Houston lifted the window and stepped inside.

He had ten more floors to go.

He took off the suit and rolled it up into a tight package, then dropped it out the window. It landed with a barely audible thump. Houston took a deep breath, drew his stun gun, and headed for the stairway.

On the landing of the sixtieth floor of the Lasser Building, David Houston paused for a moment.

"Sounds like you're out of breath," said the voice in his ear.

"You try climbing all that way sometime," Houston whispered. "I'm no superman, you know."

"Shucks," said the voice, "you've disillusioned me. What now?"

"I'm going to try to get a little information," Houston told him. "Hold on."

On the other side of the door, he could hear faint sound, as if someone were moving around, but he could hear no voices.

Carefully, he sent out a probing thought, trying to see if he could attune his mind with that of someone inside without betraying himself.

He couldn't detect anything. The sixtieth floor covered a lot of space; if whoever was inside was too far away, their thoughts would be too faint to pick up unless Houston stepped up his own power, and he didn't want to do that.

Cautiously, he reached out a hand and eased open the door.

The hallway was brightly lit, but there was no one in sight. The unaccustomed light made Houston blink for a moment before his eyes adjusted to it; the hallways and landings below had been pitch dark, forcing him to use a penlight to find his way up.

He stepped into the hallway, closing the door behind him.

Now he could hear voices. He stopped to listen. The conversation was coming from an office down the hall—if it could be called a conversation.

There would be long periods of silence, then a word or two: "But not that way." "Until tomorrow." "Vacillates."

There were three different voices.

Houston moved on down the hall, his stun gun ready. A few yards from the door, he stopped again, and, very gently, he sent out another thought-probe, searching for the minds of those within, carefully forging his way.

And, at that crucial instant, a voice spoke in his ear.

"Houston! What's going on? You haven't said a thing for two full minutes!"

"I'm all right!" Houston snapped. Only the force of long training and habit kept him from shouting the words aloud instead of keeping them to a subvocal whisper.

"All right or not," said the other, "we're coming in in seven minutes, as ordered. Meanwhile, there's a news bulletin for you; the British division has picked up another Controller—a woman named Dorrine Kent. Two in one night ought to be a pretty good bag."

For a moment, Houston's mind was a meaningless blur.

Dorrine!

And then another voice broke through his shock.

"Dear me, sir! Calm yourself! You're positively fizzing!"

Houston jerked. Standing in the doorway of the office was Norcross Lasser, with a benign smile on his face and a deadly-looking .38 automatic in his hand. Behind him stood John Sager and Loris Pederson, their faces wary.

"Please drop that stun gun, Mr. Cop."

In those few moments, Houston had regained control of himself. He realized what had happened. The interruption of his thought-probe had startled him just a little, but that little had been enough to warn the Controller.

He wondered which of the three men was the actual Controller.

He began to lower his weapon, then, suddenly, with all the force and hatred he could muster, he sent a blistering, shocking thought toward the man with the gun.

Lasser staggered as though he'd been struck. His gun wavered, and Houston fired quickly with his stun gun. At the same time, Lasser's automatic went off.

The bullet went wild, and the stun beam didn't do much better. It struck Lasser's hand, paralyzing it, but it didn't knock out Lasser.

The mental battle that ensued only took a half second, but at the speed of thought, a lot of things can happen in a half second.

Houston realized almost instantaneously that he had made a vast mistake. He had badly underestimated the enemy.

There was no need to worry, now, about which one of the men was a Controller—*all three of them were!*

As soon as Sager and Pederson realized what had happened, they leaped—mentally—into the battle. Lasser, already weakened by the unexpected mental blow from Houston, lost consciousness when the others let loose their blasts because his mind was still linked with Houston's, and he absorbed a great deal of the mental energy meant for Houston's brain.

Houston, fully warned by now, held up a denial wall which screened his mind from the worst that Sager and Pederson could put out, but he knew he couldn't hold out for long.

"Come in—*now!*" he said hoarsely into the microphone.

"Stupid swine!" Sager susurrated sibilantly.

Pederson said nothing aloud, but his brain was blazing with fear and hatred. His gun hand jerked towards a holster under his arm. Lasser was still crumpling towards the floor.

The entire action had taken less than a second.

Houston tried to fire again with his stun gun, but it required every bit of concentration he could sum up to hold off the combined mental assaults of Sager and Pederson.

But they, too, were at somewhat of a disadvantage. In order to keep all their efforts concentrated on the PD policeman, both Controllers had to refrain from putting too much attention on their bodily motions. Pederson was still fumbling for his gun, and Sager hadn't yet started for his.

Lasser barely touched the floor before his consciousness began to return. The resulting fraction of a second of mental static afforded Houston a brief respite; it disturbed Pederson just as he was getting his fingers on the butt of his weapon.

Both Controllers were focusing their mental energies on Houston's brain, and during the brief respite, Houston made one vital mental adjustment. He allowed both thought-probes to fuse in a small part of his consciousness. They went *through* him and lashed back at the two Controllers.

Both of them had had their minds tuned to Houston's, and in that instant they found they, were also attuned to each other.

The resultant of the energy was shocking to Houston, but it was infinitely worse for Sager and Pederson, since neither of them had been expecting it. Pederson, who had already been slightly distracted, got the major brunt of the force. He managed to jerk his gun free, but his brain was already lapsing into unconsciousness.

Houston's fingers tightened on his own weapon. It fired once at Lasser, who was trying to lift himself from the floor. Then it swept up and coughed again, dropping Pederson. His pistol barked once, sending a singing ricochet along the hall.

Sager, who had staggered to one side when he and Pederson had short-circuited each other, had time to get behind the protection of the office door. He couldn't close it because Lasser's and Pederson's inert forms blocked the doorway, but at least it afforded protection against Houston's stun gun.

His thought came through to Houston: *So the stupid Normals have a Controller working for them! Traitor!*

You're the traitor, Houston thought coldly. *You and your megalomaniac friends. It's madmen like you who have made telepaths hated and feared by the Normals.*

And so they should hate and fear us, came the snarling mental answer. *Within a few generations, we will have supplanted them. We will control Earth—not they.*

The exchange had only taken a fraction of a second. Houston was already charging toward the open door, hoping to get inside before Sager could reach a weapon.

You call me a traitor, Houston thought, *but you have been framing innocent Controllers, putting them into the hands of the PD Police.*

That's a lie! the reply came hotly. *We would never betray another telepath to the stupid Normals! If a telepath were so bullheaded as to get in our way, we'd dispose of him. But it would be Controller justice; we wouldn't turn him over to animals!*

In one blazing moment, Houston realized that the Controller was telling the truth!

No mental communication can be expressed properly in words. In, behind, and around each statement, other, dimmer nuances of thought gleam through. Each thought tells the receiver much more than can be put down in crude verbal symbols.

Thus, Houston already knew that Lasser, Sager, and Pederson were the three top men in a world-wide clique of megalomaniac Controllers. This was the top of the madmen's organization; these three were the *creme de la creme* of the Normal human's real enemies.

He knew that there were twelve others scattered over Earth, and he knew where and who they were. That brief exchange had brought all the information into Houston's own mind as it leaked from the minds of the others. He knew it without thinking about how he knew it.

And they were *not* the ones who had been turning the sane Controllers over to the Psychodeviant Police!

Then who was? And why?

Houston was right back where he had started.

But that brief instant of confusion was Houston's downfall. Sager instantly realized that he had delivered, inadvertently, a telling blow to Houston's mind.

Physically, Houston had been propelling himself toward the open door. At the instant of the revelation, he had been part way through it. And at that moment, Sager acted.

He slammed all his weight violently against his side of the door, knocking Houston off balance as the door swung and struck him. He went down, and Sager was on top of him before he struck the floor.

It was the weirdest battle ever fought, but its true worth could only have been detected by another telepath. It was intense and brutal.

The men fought both physically and mentally. They struggled for possession of the stun gun, at the same time hurling emotion-charged shafts of mental energy at each other's brains.

The struggle lasted less than a minute. Somehow, Sager managed to get one hand on the gun, twisting it. Houston, trying to keep it out of Sager's hand, jerked it up between them.

It coughed once, sending a beam of supersonic energy into the bodies of both men.

The effect was the same as if they had both been crowned with baseball bats.

Little pinpoints of light against a sea of darkness.

I'm cold, Houston thought. *And I'm sick.*

He couldn't tell whether his eyes were open or closed—and he didn't much care.

He tried to move his arms and legs, found he couldn't, and gave it up.

He blinked.

My eyes must be open, he thought, *if I can blink.*

Well, then, if his eyes were open, why couldn't he see anything? All he could see were the little pinpoints of light against a background of utter blackness.

Like stars, he thought.

Stars? STARS!

With a sudden rush, total awareness came back to him, and he realized with awful clarity where he was.

He was chained, spread-eagled, on an asteroid in the Penal Cluster, nearly a hundred million miles from Earth.

It was easy to piece together what had happened. He dimly remembered that he had started to wake up once before. It was a vague, confused recollection, but he knew what had taken place.

The PD Police, coming in response to his call, had found all four men unconscious from the effects of the stun beam. Naturally, all of them had been taken into custody; the PD Police had to find out which one of the men was the Controller and which the controlled. That could easily be tested by waiting until they began to wake up; the resulting mental disturbances would easily identify the telepath.

Houston could imagine the consternation that must have resulted when the PD men found that all three suspects—*and* their brother officer—were Controllers.

And now here he was—tried, convicted, and sentenced while he was unconscious—doomed to spend the rest of his life chained to a rock floating in space.

A sudden chill of terror came over him. Why wasn't he asleep? Why wasn't he under hibernene?

It's their way of being funny, came a bitter thought. *We're supposed to be under hibernene, but we're left to die, instead.*

For a moment, Houston did not realize that the thought was not his own, so well did it reflect his own bitterness. It was bad enough to have to live out one's life under the influence of the hibernation drug, but it was infinitely worse to be conscious. Under hibernene, he would have known nothing; his sleeping mind in his comatose body would never have realized what had happened to him. But this way, he would remain fully awake while his body used up the air too fast and his stomach became twisted with hunger pangs which no amount of intravenous feeding could quell. Oh, he'd live, all right—for a few months—but it would be absolute hell while he lasted. Insanity and catatonia would come long before death.

That's a nasty thought; I wish you hadn't brought it up.

That wasn't his own thought! There was someone else out here!

Hell, yes, my friend; we're all out here.

"Where are you?" Houston asked aloud, just to hear his own voice. He knew the other couldn't hear the words which echoed so hollowly inside the bubble of the spacesuit helmet, but the thought behind them would carry.

"You mean with relation to yourself?" came the answer. "I don't know. I can see several rocks around me, but I can't tell which one you're on."

Houston could tell now that the other person was talking aloud, too. So great was the illusion carried to his own brain that it almost seemed as though he could hear the voice with his ears.

"Then there are others around us?" Houston asked.

"Sure. There were three of us: a Hawaiian named Jerry Matsukuo; a girl from Bombay, Sonali Siddhartha; and myself, Juan Pedro de Cadiz. Jerry and Sonali are taking a little nap. You're the first of your group to wake up."

"My group?"

"Certainly, my friend. There are five of you; the other four must still be unconscious."

Four? That would be Lasser, Sager, Pederson, and—*and Dorrine!*

Juan Pedro de Cadiz picked up the whole thought-process easily.

"The girl—I'm sorry," he said. "But the other three—of us all, I think, they deserve this."

"Juan!" came another thought-voice. "Have our newcomers awakened?"

"Just one of them, my sweet," replied the Spaniard. "Sonali, may I present Mr. David Houston. Mr. Houston, the lovely Sonali Siddhartha."

"Juano has a habit of jumping to conclusions, David," said the girl. "He's never even seen me, and I'm sure that after three weeks of being locked in this prison whatever beauty I may have had has disappeared."

"Your thoughts are beautiful, Sonali," said Juan Pedro, "and with us, that is all that counts."

"It is written," said a third voice, "that he who disturbs the slumber of his betters will wake somebody up. You people are giving me dreams, with your ceaseless mental chatter."

"Ah!" the Spaniard said. "Mr. Matsukuo, may I—"

"I heard, Romeo, I heard," said the Hawaiian. "An ex-cop, eh? I wonder if I like you? I'll take a few thousand years to think it over; in the meantime, you may treat me as a friend."

"I'll try to live down my reputation," said Houston.

It was an odd feeling. Physically, he was alone. Around him, he could see nothing but the blackness of space and the glitter of the stars. He knew that the sun must be shining on the back of his own personal asteroid, but he couldn't see it. As far as his body was concerned, there was nothing else in the universe but a chunk of pitted rock and a set of chains.

But mentally, he felt snug and warm, safe in the security of good friends. He felt—

"David! David! Help me! Oh, David, David, David!"

It was Dorrine, coming up from her slumber. Like a crashing blare of static across the neural band, her wakening mind burst into sudden telepathic activity.

Gently, Houston sent out his thoughts, soothing her mind as he had soothed Harris's mind weeks before. And he noticed, as he did it, that the other three were with him, helping. By the time Dorrine was fully awake, she was no longer frightened or panicky.

"You're wonderful people," she thought simply, after several minutes.

"To one so beautiful, how else could we be?" asked Juan Pedro.

"Ignore him, Dorrine," said Sonali, "he tells me the same thing."

"But not in the same way, *amiga*!" the Spaniard protested. "Not in the same way. The beauty of your mind, Sonali, is like the beauty of a mountain lake, cool and serene; the beauty of Dorrine is like the beauty of the sun—warm, fiery, and brilliant."

"By my beard!" snorted Matsukuo. "Such blather!"

"I'll be willing to wager my beautiful *hacienda* in the lovely countryside of Aragon against your miserable palm-leaf *nipi* shack on Oahu that you have no beard," said Juan Pedro.

"Hah!" said Matsukuo; "that's all I need now—Castles in Spain."

It was suddenly dizzying for Houston. Here were five people, doomed to slow, painful death, talking as though there were nothing to worry about. Within minutes, each had learned to know the others almost perfectly.

It was more than just the words each used. Talking aloud helped focus the thoughts more, but at the same time, thousands of little, personal, fringe ideas were present with the main idea transmitted in words. Houston had talked telepathically to Dorrine hundreds of times, but never before had so much fine detail come through.

Why? Was there something different about space that made mental communication so much more complete?

"No, not that, I think," said Matsukuo. "I believe it is because we have lost our fear—not of death; we still fear death—but of betrayal."

That was it. They knew they were going to die, and soon. They had already been sentenced; nothing further could frighten them. Always before, on Earth, they had kept their thoughts to themselves, fearing to broadcast too much, lest the Normals find them out. The little, personal things that made a human being a living personality were kept hidden behind heavy mental walls. The suppression worked subconsciously, even when they actually wanted to communicate with another Controller.

But out here, there was nothing to fear on that score. Why should they, who were already facing death, be afraid of anything now?

So they opened up—wide. And they knew each other as no group of human beings had ever known each other. Every human being has little faults and foibles that he may be ashamed of, that he wants to keep hidden from others. But such things no longer mattered out here, where they had nothing but imminent death and the emptiness of space—and each other.

Physically, they were miserable. To be chained in one position, with very little room to move around, for three weeks, as Sonali had been, was torture. Sonali had been there longer than the others—for three days, there had been no one but herself out there in the loneliness of space.

But now, even physical discomfort meant little; it was easy to forget the body when the mind was free.

"What of the others?" Dorrine asked. "Where are the ones who were sentenced before us?"

Houston thought of Robert Harris. What had happened to the young Englishman?

"Space is big," said Juan Pedro. "Perhaps they are too far away for our thoughts to reach them—or perhaps they are already dead."

"Let's not talk of death." Sonali Siddhartha's thought was soft. "We have so many things to do."

"We will have a language session," said Juan Pedro. "*Si?*"

Matsukuo chuckled. "Good! Houston, until you've tried to learn Spanish, Hindustani, Arabic, Japanese, and French all at once, you don't know what a language session is. We—"

The Hawaiian's thought was suddenly broken off by a shrieking burst of mental static.

The effect was similar to someone dropping a handful of broken glass into an electric meat grinder right in the middle of a Bach cantata.

It was Sager, coming out of his coma.

Almost automatically, the five contacted his mind to relax him as he awoke. They touched his mind—and were repelled!

Stay out of my mind!

With almost savage fury, the still half-conscious Sager hurled thoughts of hatred and fear at the five minds who had tried to help him. They recoiled from the burst of insane emotion.

"Leave him alone," Houston thought sharply. "He's a tough fighter."

At first, Sager was terrified when he learned what had happened to him. Then the terror was mixed with a boiling, seething hatred. A hatred of the Normals who had done this to him, and an even more terrible hatred for Houston, the "traitor."

The very emptiness of space itself seemed to vibrate with the surging violence of his hatred.

"I know," Houston told him, "you'd kill me if you could. But you can't, so forget it."

Not even the power of that hatred could touch Houston, protected as he was by the combined strength of the other four sane telepaths. He was comparatively safe.

Sager snarled like a trapped animal. "You're all insane! Look at you! The four of you, siding with a man who has betrayed us to the Normals! He—"

What Sager thought of Houston couldn't be put into words, and if it could no sane person would want to repeat the mad foulness in those words.

"This is unbearable!" Sonali thought softly.

"That's not a mind," said Dorrine, "it's a sewer."

"I suggest," said Matsukuo, "that we do a little probing. Let's find out what makes this thing tick."

"Stay out of my mind!" Sager screamed. "You have no right!"

"You seemed to think you had the right to probe into the helpless minds of Normals," said Juan Pedro coldly. "We should show you how it feels."

"But they're just animals!" Sager retorted. "I am a Controller!"

"You're a madman," said Matsukuo. "And we must find out what makes you mad."

Synchronizing perfectly, five minds began to probe at the walls that Sager had built up around his personality. And as they probed, Sager retreated behind ever thicker walls, howling in hatred and anguish.

On and on went the five, needling, pressing at every weak spot, trying to break him down. Outnumbered and overpowered, it seemed as though Sager had no chance.

But his insanity was stronger than they suspected. The barriers he built were harder, more opaque, and more impenetrable than any they had ever seen. The five pushed on, anyway, but their advance slowed tremendously.

Then, mentally, there was a sudden silence.

Sager? they thought.

No answer.

"That's finished it," said Houston. "He's retreated so far behind those mental barriers that he's cut himself off completely."

"He's not dead, is he?" Dorrine asked.

"Dead?" said Juan Pedro. "Not in the sense you mean. But I think he is catatonic now; he has lost all touch with the outside. He is as though he were still drugged; he is physically helpless, and mentally blanked out."

"There's one difference," Matsukuo said analytically. "And that is that, although he has cut himself off from us and from the rest of the universe, he is still conscious in some little, walled-in compartment of his mind. He has no one there but himself—and that, I think, is damned poor company."

They waited then. When Pederson awoke, they were ready for him. His hatred took a slightly different form from Sager's, but the effect was the same.

And so were the results when the five bore down on him.

Again they waited. Lasser was next.

At first, it looked as though Lasser would go the way of Sager and Pederson, ending up as a hopelessly insane catatonic. Like his cohorts before him, Lasser retreated under the full pressure of the thought-probes of the five, building stronger and stronger walls.

But, quite suddenly, all his defenses crumbled. The mental barriers went down, shattered and dissolving. Lasser's whole mind lay bare. Instead of fighting and hating, Lasser was begging, pleading for help.

Lasser was not basically insane. His mind was twisted and warped, but beneath the outer shell was a personality that had enough internal strength to be able to admit that it was wrong and ask for help instead of retreating into oblivion.

"This one—I think we can do something with," Matsukuo's thought whispered.

Eight bodies, uncomfortable and pain-wracked, floated in space, chained to tiny asteroids that drifted slowly in their orbits under the constant pull of the sun. Two of them contained minds that were locked irrevocably within prisons of their own building, sealed off forever from external stimuli, but their suffering was the greater for all that.

The other six, chained though their limbs might be, had minds that were free—free, even, of any but the most necessary of internal limitations.

Eight bodies, chained to eight lumps of pitted rock, spun endlessly in endless space.

And then the ship came.

The flare of its atomic rocket could be seen for over an hour before it reached the Penal Cluster. The six eyed it speculatively. Although only two of them were facing the proper direction to see it with their physical eyes, the impressions of those two were easily transmitted to the other four.

"Another load of captives," whispered Juan Pedro de Cadiz. "How many this time, I wonder?"

"How long have we been here?" asked Houston, not expecting any answer.

"Who knows?" It was Lasser. "What we need out here is a clock to tell us when we'll die."

"Our oxygen tanks are our clocks," said Sonali. "And they'll notify us when the time comes."

"I do believe you morbid-minded people are developing a sense of humor," said Matsukuo, "but I'm not sure I care for the style too much."

The flare of the rocket grew brighter as the decelerating ship approached the small cluster of rocks. At last the ship itself took form, shining in the eternal blaze of the sun. When the whiteness of the rocket blaze died suddenly, the ship was only a few dozen yards from Houston's own asteroid.

And then a mental voice came into the minds of the six prisoners.

"How do you feel, Controllers?"

Only Houston recognized that thought-pattern, but his recognition was transmitted instantly to the others.

"Reinhardt!"

Hermann Reinhardt, Division Chief of the Psychodeviant Police, the one man most hated and feared by Controllers, was himself a telepath!

"Naturally," said Reinhardt. "Someone had to take control of the situation. I was the only one who was in a position to do it."

His thoughts were neither hard nor cold; it was almost as if he were one of them—except for one thing. Only the words of his thoughts came through; there were none of the fringe thoughts that the six were used to in each other.

"That's true," thought Reinhardt. "You see, we have been at this a good deal longer than you." Then he directed his thoughts at members of the crew of the spaceship, but they could still be heard by the six prisoners. "All right, men, get those people off those rocks. We have to make room for another batch."

The airlock in the side of the ship opened, and a dozen spacesuited men leaped out. The propulsion units in their hands guided them toward the prison asteroids.

"Give them all anaesthetic except Sager and Pederson," Reinhardt ordered. "They won't need it." Then, with a note of apology, "I'm sorry we'll have to anaesthetize you, but you've been in one position so long that moving you will be rather painful. We have to get you to a hospital quickly."

The minds of the six prisoners were frantically pounding questions at the PD chief, but he gave them no answer. "No; wait until you're better."

The spacesuited rescuers went to the "back" of each asteroid and injected sleep-gas into the oxygen line that ran from the tank to the spacesuit of the prisoner.

Houston could smell the sweetish, pungent odor in his helmet. Just before he blacked out, he hurled one last accusing thought at Reinhardt.

"*You're* the one who's been framing Controllers!"

"Naturally, Houston," came the answer. "How else could I get you out here?"

Houston woke up in a hospital bed. He was weak and hungry, but he felt no pain. As he came up from unconsciousness, he felt a fully awake mind guiding him out of the darkness.

It was Reinhardt.

"You're a tough man, Houston," he said mentally. "The others won't wake up for a while yet."

He was sitting on a chair next to the bed, holding a smouldering cigarette in one hand. He looked strange, somehow, and it took Houston a moment to realize that there was a smile on that broad, normally expressionless face.

Houston focussed his eyes on the man's face. "I want an explanation, Reinhardt," he said aloud. "And it better be a damned good one."

"I give you free access to my mind," Reinhardt said. "See for yourself if my method wasn't the best one."

Houston probed. The explanation, if not the best, was better than any Houston could have thought of.

When the hatred of the normal-minded people of Earth had been turned against the Controllers because of the actions of a few megalomaniacs, it had become obvious that legal steps had to be taken to prevent mob violence.

It had been Reinhardt himself who had suggested the Penal method to the UN government. At first, he had simply thought of it as a method to keep the Controllers alive until he could think of something better. But when he had discovered, by accident, what a small group of Controllers, alone in space, could do, he had set up the present machinery.

As soon as a Controller was spotted, a careful frame-up was arranged. Then, when several had been found, they were arrested in quick succession and sent to the asteroids.

Always and invariably, they had done what Houston's group had done—the sane or potentially sane ones had improved themselves tremendously, while the megalomaniacs had lapsed into catatonia.

"Why couldn't it be done on Earth?" Houston asked.

"We tried it," Reinhardt said. "It didn't work. Safe, on Earth, surrounded by Normals, a Controller still feels the hatred around him. He can't open his mind completely. Only the certain knowledge of impending death, and a complete freedom from the hatred of Normals can free the mind.

"And that's why you couldn't be told beforehand; if you knew you were going to be rescued, you wouldn't open up."

Houston nodded. It made sense. "Where are we now?" he asked.

"Antarctica," said Reinhardt. "We've built an outpost here—almost self-sufficient. When you're in better shape physically, I'll show you around."

"Do you mean that everyone who's been arrested is here, in Antarctica?"

Reinhardt laughed. "No, not by a long shot. Most of us are back out in civilization, searching for new, undiscovered Controllers, so that we can frame them. And, of course, some of us—the insane ones—have died. They will themselves to die when the going gets too tough."

"Searching for recruits? Then the Group that Dorrine was working for was—"

Reinhardt shook his head. "No. They were going about it the wrong way, just as you thought. We picked up the whole lot of them last week; they're occupying the asteroids now."

"What do you do with the insane catatonics?"

"Put them under hibernene and keep them alive. We hope, someday, to figure out a method of restoring their sanity. Until then, let them sleep."

Houston narrowed his eyes. "How long have you known I was a Controller, Reinhardt?"

The Prussian smiled. "Ever since you first tried to probe me. Fortunately, my training enabled me to put up a shield that you couldn't penetrate; I seemed like a Normal to you.

"I kept you on because I knew you'd be useful in cracking Lasser and his gang when the time came. No one else could have done what you did that night."

"Thanks," Houston said sincerely. "What's going to happen now? After I get well, I mean."

"You'll do what the others have done. A little plastic surgery to change your face a trifle, a little record-juggling to give you a new identity, and you'll be ready to go back to work for the PD Police.

"If anyone recognizes you, it's easy to take over their minds just long enough to make them forget. We allow that much Controlling."

"And then what?" Houston wanted to know. "What happens in the long run?"

"In a way," said Reinhardt, "your friend Sager was right. The Controllers will eventually become the rulers of Earth. But not by force or trickery. We must just bide our time. More and more of us are being born all the time; the Normals are becoming fewer and fewer. Within a century, we will outnumber them—we will be the Normals, not they.

"But they'll never know what's going on. The last Normal will die without ever knowing that he is in a world of telepaths.

"By the time that comes about, we'll no longer need the Penal Cluster, since Controllers will be born into a world where there is no fear of non-telepaths."

"I wonder," Houston mused, "I wonder how this ability came about. Why is the human race acquiring telepathy so suddenly?"

Reinhardt shrugged. "I can give you many explanations—atomic radiation, cosmic rays, natural evolution. But none of them really explains it. They just make it easier to live with.

"I think something similar must have happened a few hundred thousand years ago, when Cro-Magnon man, our own ancestors, first developed true intelligence instead of the pseudo-intelligence, the highly developed instincts, of the Neanderthals and other para-men.

"Within a relatively short time, the para-men had died out, leaving the Cro-Magnon, with his true intelligence, to rule Earth."

Reinhardt stood up. "Why is it happening? We don't know. Maybe we never will know, any more than we know why Man developed intelligence." He shrugged. "Perhaps the only explanation we'll ever have is to call it the Will of God and let it go at that."

"Maybe that's the best explanation, after all," Houston said.

"Perhaps. Who knows?" Reinhardt crushed his cigarette out in a tray. "I'll go now, and let you get some rest. And don't worry; I'll have you notified as soon as Dorrine starts to come out of it."

"Thanks—Chief," Houston said as Reinhardt left the room.

David Houston lay back in his bed and closed his eyes.

For the first time in his life, he felt completely at peace—with himself, and with the Universe